Totally into Fishing and Hunting

Ice Fishing

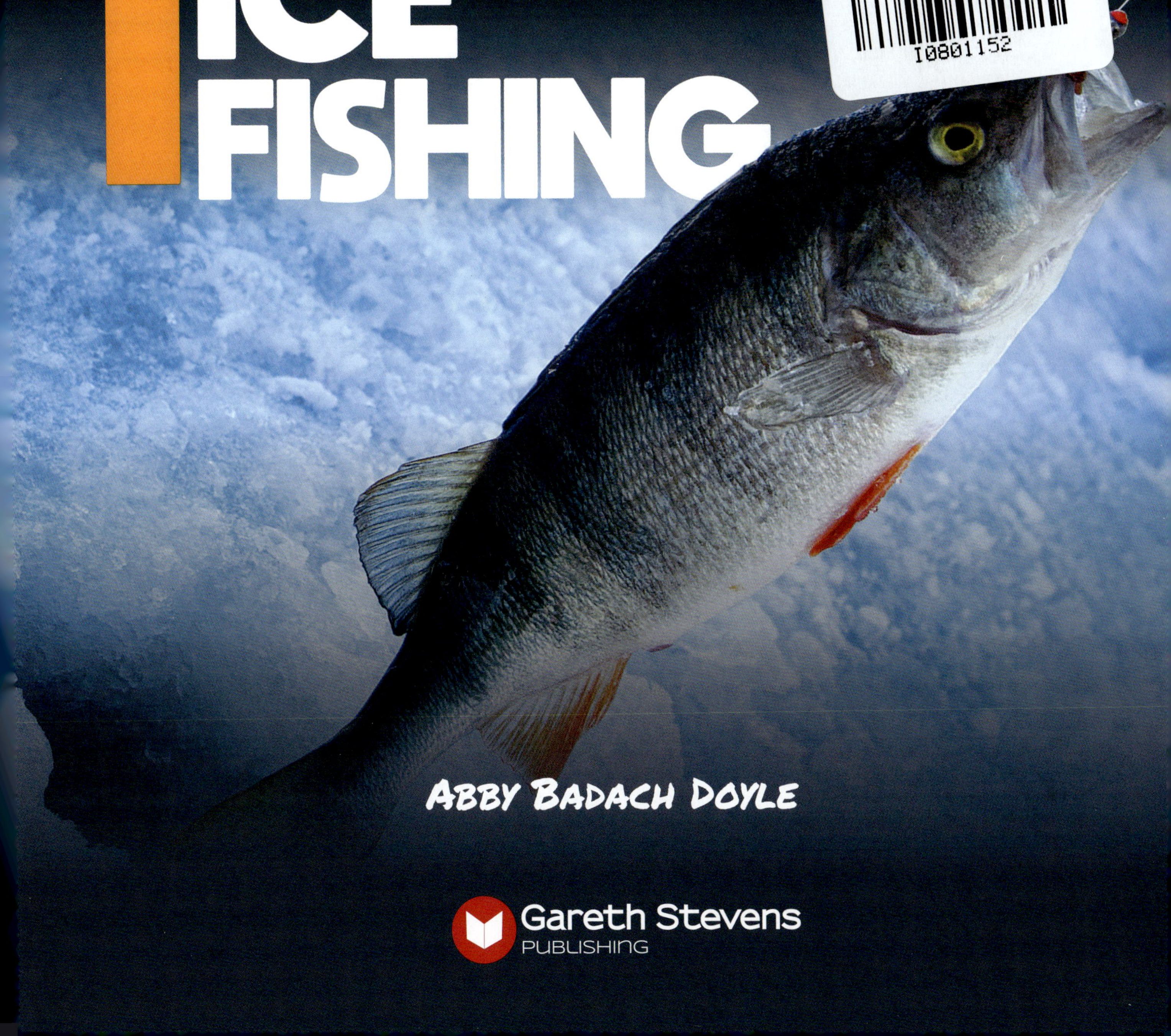

Abby Badach Doyle

Gareth Stevens Publishing

Please visit our website, www.garethstevens.com. For a free color catalog of all our high-quality books, call toll free 1-800-542-2595 or fax 1-877-542-2596.

Library of Congress Cataloging-in-Publication Data
Names: Doyle, Abby Badach, author.
Title: Ice fishing / Abby Badach Doyle.
Description: New York : Gareth Stevens Publishing, [2023] | Series: Totally into fishing and hunting | Includes index.
Identifiers: LCCN 2022003861 (print) | LCCN 2022003862 (ebook) | ISBN 9781538280089 (set) | ISBN 9781538280096 (library binding) | ISBN 9781538280072 (paperback) | ISBN 9781538280102 (ebook)
Subjects: LCSH: Ice fishing–Juvenile literature.
Classification: LCC SH455.45 .D69 2023 (print) | LCC SH455.45 (ebook) | DDC 799.12/2–dc23/eng/20220203
LC record available at https://lccn.loc.gov/2022003861
LC ebook record available at https://lccn.loc.gov/2022003862

Portions of this work were originally authored by Nia Kennedy and published as *We're Going Ice Fishing*. All new material in this edition is authored by Abby Badach Doyle.

Published in 2023 by
Gareth Stevens Publishing
29 East 21st Street
New York, NY 10010

Editor: Abby Badach Doyle
Designer: Michael Flynn

Photo credits: Cover, pp. 1 Maximillian cabinet/Shutterstock.com; series background (camo) Alexvectors/Shutterstock.com; p. 5 GROGL/Shutterstock.com; p. 6 Fedbul/Shutterstock.com; p. 7 Mataliia Korzhenevska/Shutterstock.com; p. 9 PRESSLAB/Shutterstock.com; p. 10 style-photo/Shutterstock.com; p. 11 Stephen Mcsweeny/Shutterstock.com; p. 13 Dan Thornberg/Shutterstock.com; p. 15 Stephen Mcsweeny/Shutterstock.com; p. 16 alexkich/Shutterstock.com; p. 17 Justine Wild/Shutterstock.com; p. 19 Aleron Val/Shutterstock.com; p. 20 ilmarinfoto/Shutterstock.com; p. 21 ver0nicka/Shutterstock.com; p. 23 Ramon Cliff/Shutterstock.com; p. 24 Derek Robbins/Shutterstock.com; p. 25 Joseph Kreiss/Shutterstock.com; p. 27 smspsy/Shutterstock.com; p. 29 Rocksweeper/Shutterstock.com.

Printed in the United States of America

A NOTE TO READERS
Always talk with a parent or teacher before proceeding with any of the activities found in this book. Some activities require adult supervision.

A NOTE TO PARENTS AND TEACHERS
This book was written to be informative and entertaining. Some of the activities in this book require adult supervision. Please talk with your child or student before allowing them to proceed with any hunting activities. The author and publisher specifically disclaim any liability for injury or damages that may result from use of information in this book.

Some of the images in this book illustrate individuals who are models. The depictions do not imply actual situations or events.

CPSIA compliance information: Batch #CSGS23: For further information contact Gareth Stevens, New York, New York at 1-800-542-2595.

CONTENTS

WORDS IN THE GLOSSARY APPEAR IN BOLD TYPE THE FIRST TIME THEY ARE USED IN THE TEXT.

A COOL HOBBY

Fishing isn't just something to enjoy during the warm months. You can fish during the winter too . . . if you don't mind bundling up! The **temperature** has to be very cold to go ice fishing.

This allows lakes, ponds, and other bodies of water to grow a thick layer of ice. Ice fishing is popular around the world in places with very cold winters. In North America, popular ice fishing spots are found in Canada and Upper Midwest states such as Minnesota, Michigan, and Wisconsin.

KNOW THE FACTS!

A species is a group of animals or plants that are the same kind. Popular fish species for ice fishing include yellow perch, walleye, pike, trout, crappie, and sunfish.

Millions of people go ice fishing every year in the United States.

HOW DO FISH SURVIVE?

Most fish are cold-blooded, which means their body temperature is the same as their surroundings. Freshwater fish have special **adaptations** to help them survive the icy winters. Some fish slow their breathing and heart rate. They also swim around a lot less. They don't need as much food or **oxygen** during this time.

Yellow perch, a popular ice fishing catch, travel in schools.

Many fish move to the deepest parts of lakes, ponds, or rivers. This is because the water is warmer there. They stay in schools, or groups, to stay warm.

KNOW THE FACTS!

Fresh water is water that has almost no salt in it. Oceans are made of salt water. Salt water freezes at lower temperatures than fresh water. Frozen ocean water is only found in very cold places such as the North and South Poles.

GET READY

First, learn your state's rules about fishing. Each state has different rules for where you are allowed to fish, how many you can catch, and how big the fish must be to keep. You might also need to buy a fishing **license**.

When you're still learning, it's best to go fishing with a trusted adult or a group. For safety, never go ice fishing alone. Remember to check the weather. Very cold temperatures or heavy snow can make it unpleasant to spend time outside.

KNOW THE FACTS!

Most times, children don't need a fishing license if they are fishing with a licensed adult. Most states allow you to get a fishing license when you reach age 16 or so.

A sled can help you carry your ice fishing gear.

WHAT TO WEAR

Dressing in layers can help you stay warm. Layering is when you wear one piece of clothing over another. You can remove a layer if you get too hot. Don't forget warm socks, mittens or gloves, a scarf, and a warm hat.

KNOW THE FACTS!

Some people bring other footwear, such as ice skates or snowshoes, to go ice fishing. These can make it easier to move around a frozen or snow-covered lake.

Sunglasses can guard your eyes from the bright and harmful sun that **reflects** off snow.

Sturdy boots will keep you from slipping on the ice. Bibs, or special overalls made to keep you warm up to the chest, can have reflective stripes for added safety. This helps other people see you if it gets dark.

ICE SAFETY

Ice can be thick in one spot and much thinner nearby. Thin ice is dangerous, or unsafe. You could get very hurt or drown if you fall into the freezing water underneath. The first and most important thing to do before ice fishing is to check the thickness of the ice.

Ice must be at least 4 inches (10.2 cm) thick for a person to walk on it. Some people drive snowmobiles, **ATVs**, cars, or trucks to go ice fishing. The ice must be much thicker to support **vehicles**.

KNOW THE FACTS!

New ice is stronger and safer than old ice. New ice is clear or blueish in color. Old ice looks cloudy or gray.

Ice: How Thick Is Safe?

at least 4 inches (10.2 cm)
One person walking or ice fishing

around 5–7 inches (12.7–17.8 cm)
Snowmobile or ATV

around 8–12 inches (20.3–30.5 cm)
Small car

around 12–15 inches (30.5–38.1 cm)
Medium pickup truck

No ice ever is 100 percent safe. Check the thickness every 150 feet (45.7 m).

CUTTING INTO THE ICE

Once you have found thick ice, the next step is to drill a hole. A special drill called an ice auger is the perfect tool to use for the job. The hole should not be more than 12 inches (30.5 cm) across.

Some augers are turned by hand. Others are powered by gas and a motor, or engine. With either type, it's important to have steady footing before you start drilling. After you are done drilling, clear the slush away from the hole to keep from slipping.

KNOW THE FACTS!

It's possible to make a hole without an ice auger, but it takes much more effort. Some people use a spud bar. This is a sturdy bar with a flat, sharp end.

A hole 5 to 8 inches (12.7 to 20.3 cm) across can fit most small- and medium-sized fish.

ROD AND REEL

A popular way to ice fish is with a rod and **reel**. The rods used in ice fishing are much shorter than those used in warm-weather fishing. That's because you don't have to cast, or throw your line far out into the water. You just bait your hook and drop it down the hole!

To catch big fish like this northern pike, you need a heavier fishing rod.

Ice fishing rods are usually around 24 to 36 inches (61 to 91.4 cm) long. They are usually lightweight. This makes it easier to feel when fish bite.

Know the Facts!

If you catch a fish, you can keep it to eat if it's big enough. Or you can gently return it to the water. This is known as catch-and-**release** fishing.

OTHER FISHING METHODS

Other methods of ice fishing don't require a rod and reel. Tip-up fishing involves flags that connect to your fishing line. You set up your line in the water with the flag pointing down toward the surface of the ice. When a fish bites, the flag will flip up.

Ice-fishing spears are long poles with several sharp points on the end. You must pay close attention to your fishing hole when spearfishing. When a fish swims into view, you stick it with your spear.

KNOW THE FACTS!

Spearfishing is one of the oldest methods of ice fishing. Native Peoples, such as the Ojibwa and Inuit, have fished this way for more than 1,000 years.

Some people drill multiple holes to fish using tip-up flags.

STAY SAFE

It's easy to get lost on a large, snow-covered lake. Before you go out, ask other fishers or bait shop owners where the safest ice is. If you bring a cell phone or radio, keep it dry in a plastic bag.

Bring a loud safety whistle. You can use it to get the attention of other people if you need help. Have a rope and **life preserver** nearby in case you or someone you're with falls in the water.

LIFE PRESERVER

Ice sometimes makes a cracking noise. This is normal when the temperature changes.

SEEK SHELTER

Many people go ice fishing out in the open. They sit on a stool or a bucket. This works best on warmer days with no wind. Other people prefer to use a shelter. A shelter is a covering that protects you from the wind, snow, and cold.

Some shelters, such as pop-up or fold-over shelters, are moveable. They are made of lightweight fabric, a kind of cloth, and are easy to set up and tear down. Some even have built-in sleds to make them easier to tow.

KNOW THE FACTS!

Other shelters are made of plywood. Plywood is easily found at home improvement stores or lumberyards. Plywood is a strong board made of thin pieces of wood, held together by glue.

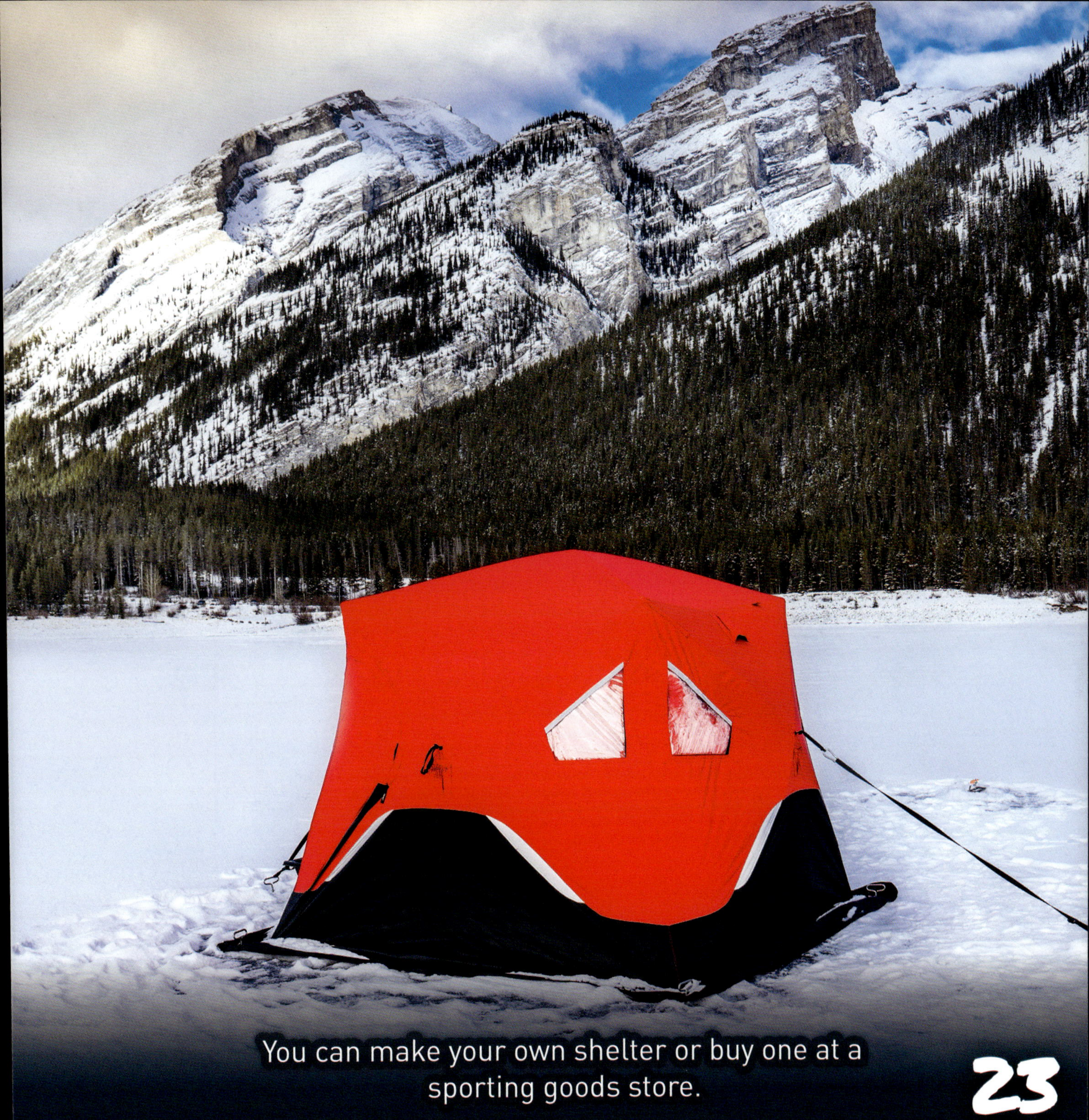

You can make your own shelter or buy one at a sporting goods store.

ICE TOWNS

Other ice fishing shelters are meant to stay up for the entire winter. They look like tiny houses or campers. Some even have furniture, electricity, kitchens, and bathrooms!

Sometimes there are so many shelters on the ice that it looks like a tiny village. If you'll be out fishing all day, bring snacks and drinks with you. Some people like to bring a radio or small TV to pass the time. Others prefer to **focus** on the fish!

KNOW THE FACTS!

People have many names for ice fishing shelters. They are also known as ice houses, bob-houses, dark houses, ice shacks, or shanties. A shanty is a name for a small, simple building.

A small heater can give you comfort in your shelter too.

FISH FINDERS

You don't need fancy gear to go ice fishing. But if you want to use **technology**, there are plenty of options. Underwater cameras can help take clear pictures of what's under the ice.

Sonar instruments can also be useful. A sonar device, or tool, gives off sound waves and picks them up again after they bounce off things underwater. Sonar is used to figure out how deep the water is. It's also used to see where weeds and

If you use a cell phone, tablet, or radio while fishing, bring an easy-to-carry charger so your devices don't run out of power.

GET HOOKED!

Ice fishing is a different adventure than regular fishing. It feels good to be active outside during the cold, dark winter months. Out on the ice, you can reach fishing spots you couldn't walk to in the summer.

You can ice fish close to home, or travel to make memories in a new place. Some people make fishing trips a party with friends and family. Others enjoy the simple quiet of being in nature. Give ice fishing a try. You might get hooked!

KNOW THE FACTS!

When you're done, don't forget to clean up your trash. Leftover hooks or fishing line can harm birds, fish, and other wildlife.

Some fishers say fish taste better when you catch them from icy waters.

GLOSSARY

adaptation: a change in a type of animal that makes it better able to live in its surroundings

ATV: the abbreviation for all-terrain vehicle. This is a vehicle made to go over uneven, varied surfaces in addition to roads

focus: to direct your attention or effort to something specific

license: a piece of paper that allows someone to do something

life preserver: a tool that floats that can save someone from drowning

oxygen: a colorless, odorless gas that many animals, including people, need to breathe

reel: a tool that holds fishing line and connects to a fishing rod

reflect: to give back light

release: to set something free

technology: using science, engineering, and other industries to invent useful tools or to solve problems

temperature: how hot or cold something is

vehicle: an object that moves people from one place to another

FOR MORE INFORMATION

BOOKS

Allen, Bob. *Walter: Under the Ice.* Minneapolis, MN: Itasca Books, 2020.

Coulson, Art. *Fishing on Thin Ice.* North Mankato, MN: Capstone, 2022.

WEBSITES

Does Salt Water Freeze?
wonderopolis.org/wonder/does-salt-water-freeze
Understand how salt water freezes and try it for yourself at home.

Learn to Ice Fish
dnr.state.mn.us/gofishing/learn-ice-fish.html
Watch videos and learn tips to get ready for your next ice fishing trip.

Publisher's note to educators and parents: Our editors have carefully reviewed these websites to ensure that they are suitable for students. Many websites change frequently, however, and we cannot guarantee that a site's future contents will continue to meet our high standards of quality and educational value. Be advised that students should be closely supervised whenever they access the internet.

INDEX